"THE AI VERDICT"

The Future of Law and Justice in India

NARAYANA SWAMY .R

Advocate Karnataka

NARAYANASWAMY.R

Copyright © 2024 All rights reserved.

No part of this publication may be reproduced, stored in a retrieval system, or transmitted in any form or by any means—electronic, mechanical, photocopying, recording, or otherwise—without the prior written permission of the publisher

ACKNOWLEDGEMENTS & DEDICATION

Writing **"THE AI VERDICT: The Future of Law and Justice in India"** has been an incredible journey that would not have been possible without the unwavering support and inspiration of those around me. This book is a product of dedication, countless hours of research, and the encouragement of family and mentors who have stood by me.

First and foremost, I wish to express my heartfelt gratitude to my late father, **Rangappa M.K.**, whose values of integrity, perseverance, and a relentless pursuit of knowledge have been the guiding principles in my life. Though he is no longer with us, his legacy continues to inspire every step I take.

To my mother, **Lakshminarasamma,** whose boundless love, wisdom, and resilience have shaped my character and instilled in me the importance of compassion and fairness—qualities that I hope to reflect in this work—I owe a deep debt of gratitude. Her encouragement has been a source of strength and motivation throughout this endeavor.

NARAYANASWAMY.R

My brother, **Shivashankara**, has been a constant pillar of support and a source of insightful conversations that have enriched my understanding and broadened my perspectives. His belief in my work has given me the confidence to push forward even when the path was challenging.

To my beloved wife, **Gowthami** M.N., whose patience, unwavering support, and understanding have been invaluable throughout the writing of this book—thank you for being my partner in every sense. Your encouragement has been the light that guided me during late nights and long days.

I would also like to acknowledge the contributions of my senior colleagues, mentors, and friends whose insights have helped shape the direction of this book. My appreciation extends to my readers, who I hope will find value and inspiration in these pages.

Preface

"THE AI VERDICT: The Future of Law and Justice in India" is a deep exploration of the evolving relationship between advanced technology and the judiciary system. This book is inspired by the potential and the challenges that the Indian legal system faces as it stands at the crossroads of tradition and innovation. With one of the most intricate judicial structures in the world and a case backlog that stretches into millions, India is ripe for reform and modernization. The introduction of Artificial Intelligence (AI) into this landscape represents a significant turning point.

AI's transformative power lies in its ability to handle massive data, enhance case management, and automate routine tasks that have traditionally consumed substantial judicial resources. From predictive analytics that help anticipate case outcomes to AI-driven legal research that can streamline the preparation process for advocates, the possibilities are vast. However, these advancements come with inherent challenges that extend beyond technology, touching on issues of fairness, bias, and the safeguarding of rights.

The objective of this book is not only to showcase the potential benefits of AI in enhancing the efficiency of the judiciary but also to acknowledge and dissect the regulatory, ethical, and social considerations that accompany such a paradigm shift. While AI can be an enabler of faster and more precise legal processes, it must be implemented thoughtfully, ensuring it complements rather than compromises the principles of justice.

Throughout "THE AI VERDICT," we delve into real-world examples, existing AI tools, and pilot projects that have shown promise in India and beyond. We also examine the global practices that can serve as a template for India's journey and explore the steps that policymakers, legal practitioners, and technologists must take to realize an AI-augmented judiciary that upholds the highest standards of fairness, impartiality, and human rights.

This book aims to be a guide and a resource for those invested in the future of law and justice in India—advocates, judges, legal scholars, policymakers, and anyone curious about how AI is poised to transform the legal world. As the legal fraternity navigates this unprecedented shift, "THE AI VERDICT" seeks to foster informed dialogue, encourage responsible adoption, and inspire confidence in an AI-enabled judicial future that remains true to the essence of justice.

In this journey through innovation and reform, we invite you to consider not just how AI can be used in the judiciary, but how it should be used to uphold the fundamental ideals of fairness, equity, and access to justice for all.

TABLE OF CONTENTS

INTRODUCTION..8
CURRENT CHALLENGES IN THE INDIAN JUDICIARY SYSTEM
....10
ROLE OF AI IN THE INDIAN JUDICIARY.......................................18
KEY AI TOOLS AND PROJECTS IN INDIA 25
BENEFITS OF AI IN THE INDIAN JUDICIARY 31
CHALLENGES AND CONCERNS..38
FUTURE PROSPECTS 44
REGULATORY AND ETHICAL CONSIDERATIONS 51
CONCLUSION 57

INTRODUCTION

The integration of Artificial Intelligence (AI) into the judiciary has become a topic of increasing global significance, with its potential to revolutionize how legal systems operate. In India, a country known for its robust and complex judicial landscape, the adoption of AI presents an unprecedented opportunity to address some of its most pressing challenges, such as case backlogs, procedural inefficiencies, and access to justice. The balance between modern technology and age-old legal traditions offers a unique landscape for innovation and reform.

The Indian judiciary, characterized by its layered structure, intricate processes, and high case volume, faces persistent challenges that have often strained its ability to deliver timely and effective justice. AI, with its capabilities in data analysis, language processing, and automation, has emerged as a potential solution that can bolster these operations. By automating routine tasks, enhancing case management, and providing sophisticated data-driven insights, AI could redefine the efficiency and accessibility of the courts.

However, the implementation of AI in judicial processes is not without concerns. Ethical considerations such as fairness, accountability, and transparency must be addressed to ensure that AI supports justice rather than compromises it. Regulatory frameworks and oversight are essential to guide this integration responsibly, maintaining the foundational principles of justice and human rights.

This exploration covers several critical aspects of AI's role in the Indian judiciary, from current projects and tools to the benefits and challenges of its application. It also delves into the future prospects of AI in law and the essential regulatory and ethical considerations required for its responsible use. The goal is to provide a comprehensive understanding of how AI can influence India's legal landscape and contribute to a more efficient and just system.

CURRENT CHALLENGES IN THE INDIAN JUDICIARY SYSTEM

The Indian judiciary system, while integral to the upholding of democracy and the rule of law, faces a multitude of challenges that impede its ability to deliver timely and effective justice. These challenges range from logistical and operational issues to systemic inefficiencies and socio-political complexities. Here is an in-depth examination of the current challenges confronting the Indian judiciary system:

1. Overwhelming Case Backlog

One of the most significant challenges facing the Indian judiciary is the immense backlog of cases. As of recent data:

- Pending Cases: Over 51 million cases are pending across various courts in India, including the Supreme Court, high courts, and subordinate courts. The district and subordinate courts account for the majority of these pending cases.

- Delays in Justice: The backlog leads to prolonged trials and delays in delivering justice, which goes against the fundamental principle of "justice delayed is justice denied." Litigants often have to wait years, sometimes decades, for the resolution of their cases.

- Impact on Public Trust: This delay erodes public confidence in the judicial system, making people less inclined to seek legal remedies for their grievances.

2. Inadequate Judge Strength

The shortage of judges is another critical problem:

- Judge-to-Population Ratio: India has one of the lowest judge-to-population ratios in the world. While the Law Commission of India recommended increasing this ratio to 50 judges per million people, the actual number is significantly lower, hovering around 20 judges per million.

- Vacant Positions: High vacancy rates in courts further exacerbate the problem. For instance, high courts and lower courts often operate with unfilled judicial positions, placing additional strain on the existing judges.

- Recruitment and Appointments: Delays in the appointment of judges due to bureaucratic and procedural inefficiencies contribute to the problem. The process of recruiting judges is often slow, involving multiple layers of scrutiny and approvals.

3. Lack of Infrastructure

Insufficient infrastructure is a notable issue affecting the judiciary's operational efficiency:

- Court Facilities: Many courts, especially those in rural and semi-urban areas, lack basic facilities such as proper seating, clean restrooms, and adequate space for both litigants and legal professionals.

- Technological Shortcomings: While there has been progress in digitizing court records and implementing e-courts, the pace is uneven. Many courts are still reliant on outdated manual processes, hindering efficiency and the swift dispensation of justice.

- Budget Constraints: The judiciary is often underfunded, receiving only a small portion of the overall government budget. This limits the scope for modernizing infrastructure and integrating technology to expedite processes.

4. Complex Legal Procedures

The complexity of legal processes contributes to delays and inefficiencies:

- Lengthy Procedures: The Indian legal system is characterized by intricate and formalistic procedures that can be difficult to navigate, especially for laypeople. Filing and admitting cases, serving notices, and gathering evidence are often slow and cumbersome.

- Procedural Adjournments: Adjournments are frequently granted, sometimes without substantial justification. This culture of adjournments, often sought by lawyers and accepted by courts, prolongs cases and adds to the backlog.

- Multiplicity of Laws: The coexistence of central, state, and local laws can create confusion and lead to legal disputes. Ensuring consistency in the application and interpretation of laws can be challenging for judges.

5. Access to Justice

Accessibility to justice remains uneven across the country:

- Urban vs. Rural Divide: While metropolitan areas may have better-equipped courts and access to experienced legal practitioners, rural areas face significant challenges, including limited legal resources and awareness.

- Economic Barriers: High legal fees and the cost of litigation can deter economically weaker sections from pursuing justice. Although initiatives like legal aid services exist, their reach and effectiveness can be inconsistent.

- Legal Literacy: A significant portion of the population is not well-informed about their legal rights and the procedures for seeking redress. This lack of legal literacy further discourages individuals from approaching the judiciary for resolution.

6. Case Complexity and Volume

The sheer volume and diversity of cases, ranging from civil disputes to complex criminal matters, present another challenge:

- Criminal Cases: The number of criminal cases, especially those involving heinous crimes, has increased. These cases often require thorough investigation, which can delay trial processes.

- Civil Litigation: Civil disputes, including property and family matters, tend to drag on for years due to multiple hearings, adjournments, and appeals.

- Specialized Cases: The rise in specialized areas such as cybercrime, environmental law, and intellectual property rights has introduced complexities that require specific expertise and additional time for adjudication.

7. Judicial Accountability and Transparency

Concerns regarding judicial accountability and transparency affect public trust:

- Judicial Conduct: While the judiciary is regarded as independent, instances of corruption or perceived bias can damage its reputation. Mechanisms for addressing complaints against judges are often seen as inadequate or lacking transparency.

- Decision-Making Process: The process by which judgments are reached, especially in high-profile cases, may be seen as opaque. Greater transparency in court proceedings and judicial reasoning can enhance public trust.

- Appellate System: The multi-tiered appellate system, while intended to ensure thorough review, can contribute to delays and inconsistent outcomes. Litigants may exploit this system to prolong litigation.

8. Impact of Socio-Political Factors

The judiciary does not operate in a vacuum; socio-political factors can also play a significant role:

- Judicial Independence: While the judiciary is constitutionally independent, it may come under pressure from political influences in certain high-profile cases or appointments.

- Public Interest Litigation (PIL): While PILs have been a tool for social change, their misuse or excessive filing can burden the judiciary and divert attention from more pressing cases.

- High-Profile Cases: The focus on high-profile cases can sometimes divert judicial resources and attention, leading to delays in addressing less-publicized but equally important cases.

9. Emerging Legal Areas and Preparedness

The fast-paced evolution of technology and global trends has introduced new areas of law that require judicial understanding:

- Cyber Law: The rapid growth of digital technology has led to a surge in cases related to cybercrime, data privacy, and online fraud. Judges need to be trained in these emerging areas to effectively handle such cases.

- Environmental Law: Climate change and environmental concerns have brought about an increase in cases related to environmental regulation and compliance. Ensuring judges have the requisite knowledge in these areas is crucial.

- Artificial Intelligence (AI) and Automation: As AI continues to make its way into legal practices, the judiciary must stay informed about the ethical and procedural implications of using AI in legal contexts.

10. Reforms and Suggestions for Improvement

To overcome these challenges, a series of reforms and policy changes are needed:

- Filling Judicial Vacancies: Accelerating the recruitment process for judges can help reduce the burden on the current judicial workforce.

- Infrastructure Development: Greater investment in infrastructure, including better facilities and technology integration, can enhance the judiciary's functioning.

- Simplification of Procedures: Streamlining and simplifying legal procedures can make the process more efficient and accessible.

- Enhanced Legal Aid Services: Strengthening legal aid mechanisms can help bridge the gap for economically weaker sections and improve access to justice.

- Training and Capacity Building: Continuous training programs for judges on new laws, technology, and emerging legal trends can prepare them for the challenges of modern jurisprudence.

- Promoting Alternative Dispute Resolution (ADR): Encouraging out-of-court settlements through mediation, arbitration, and conciliation can reduce the caseload and expedite dispute resolution.

- Technological Advancements: Expanding the e-court initiative, enhancing case management systems, and adopting AI tools can improve the efficiency of case handling.

Conclusion

The challenges faced by the Indian judiciary are multifaceted, requiring a comprehensive and well-coordinated approach to address them effectively. While significant strides have been made in some areas, persistent issues such as case backlogs, infrastructure deficits, and complex procedures need sustained attention and strategic reforms. The implementation of innovative solutions, combined with a commitment to transparency, accountability, and accessibility, can transform the judiciary into a more robust and effective pillar of democracy.

NARAYANASWAMY.R

ROLE OF AI IN THE INDIAN JUDICIARY

The role of Artificial Intelligence (AI) in the Indian judiciary is becoming increasingly prominent as the system looks for innovative solutions to address longstanding challenges such as case backlogs, procedural inefficiencies, and limited resources. AI, with its capabilities to process large amounts of data, automate repetitive tasks, and provide predictive analytics, holds the potential to transform how justice is administered. Below is an in-depth analysis of how AI is contributing to and reshaping the Indian judiciary.

1. Enhancing Case Management and Allocation

AI tools can play a crucial role in the management and allocation of cases, which is fundamental for streamlining judicial processes:

- Case Distribution: AI can be used to analyze a judge's current workload, expertise, and the complexity of pending cases to help allocate new cases efficiently. This ensures that cases are distributed equitably and matched to judges based on their specialization, improving overall productivity.

- Prioritization of Cases: Through machine learning algorithms, AI systems can evaluate case characteristics and prioritize those that require urgent attention, such as cases involving fundamental rights, high-profile criminal trials, or time-sensitive civil matters.

2. AI-Driven Legal Research and Document Analysis

Legal research is one of the most time-consuming tasks for judges and lawyers. AI can expedite this process significantly:

- Automated Legal Research: Advanced AI systems are capable of parsing through vast legal databases, past judgments, statutes, and case laws in seconds. This reduces the time spent on manual searches and ensures that legal professionals have access to the most relevant precedents and materials.

- Document Analysis and Summarization: AI-powered tools can assist in scanning and summarizing lengthy legal documents, including petitions, contracts, and evidence files. This can help identify key points, inconsistencies, or relevant citations, saving both judges and lawyers substantial time.

- Error Detection: AI can aid in spotting errors or inconsistencies in legal documents, ensuring higher accuracy and reliability of court submissions.

3. Predictive Analytics for Case Outcomes

AI systems can be trained to predict the potential outcomes of cases based on historical data:

- Data-Driven Insights: By analyzing previous judgments, legal strategies, and outcomes, AI can provide insights into how a particular case may unfold. This allows litigants and lawyers to make more informed decisions, assess risks, and potentially avoid lengthy litigation if a favorable out-of-court settlement is deemed more appropriate.

- Assisting Legal Strategy: Lawyers can use predictive analytics to shape their legal strategies by understanding patterns in judicial behavior and probable case trajectories. This insight is particularly beneficial in complex cases that involve multifaceted legal arguments.

4. Supporting Judges with AI-Powered Decision-Making Tools

While AI cannot replace human judgment in legal proceedings, it can serve as a valuable assistant:

- Research Assistance: AI-powered platforms like SUPACE (Supreme Court Portal for Assistance in Court Efficiency) help judges by compiling relevant case laws and precedents that pertain to specific legal questions being considered. This improves the speed and thoroughness of judicial research.

- Reducing Cognitive Load: By automating routine research and data analysis, AI allows judges to focus more on nuanced legal interpretation and judgment writing, reducing their cognitive load and risk of oversight.

5. Improving Accessibility and Inclusivity

One of the major challenges in the Indian judiciary is ensuring that justice is accessible to all, especially in rural and underprivileged areas:

- Language Translation: India is a linguistically diverse country, and court proceedings or legal documents are often available only in specific regional languages. AI-powered translation tools can bridge language barriers, translating judgments, legal documents, and court proceedings into multiple languages for better comprehension and access.

- Speech Recognition and Voice-to-Text: AI technologies that convert speech to text can help transcribe court proceedings, ensuring that written records are kept up to date and making it easier for individuals with hearing or visual impairments to access information.

6. Facilitating Virtual Courts and Online Dispute Resolution (ODR)

The COVID-19 pandemic accelerated the shift towards virtual court proceedings, where AI plays an essential supportive role:

- Streamlining Virtual Hearings: AI can assist in scheduling virtual court sessions, notifying parties, managing case files, and tracking hearing times, making virtual courtrooms more efficient.

- Online Dispute Resolution: AI-driven ODR platforms facilitate the resolution of disputes without the need for physical court appearances. AI can guide parties through negotiation and mediation by analyzing positions, proposing fair outcomes, and managing case files and documentation.

- Automated Docket Management: AI systems can maintain case dockets automatically, updating them with hearing dates, orders, and other relevant case progress details, ensuring that all stakeholders have real-time access to the case status.

7. Use of AI in Legal Education and Training

AI can contribute significantly to the education and training of judges and legal practitioners:

- Interactive Learning Tools: AI-powered tools can simulate court cases, presenting hypothetical scenarios that help train judges and lawyers on how to navigate different legal situations.

- Legal Analysis Training: AI can be used to develop training modules that help legal professionals understand how to analyze judgments, identify key legal principles, and apply them effectively in real-world scenarios.

8. AI for Legal Assistance to the Public

Providing the general public with access to basic legal information and guidance is crucial:

- Chatbots and Legal Assistants: AI chatbots equipped with natural language processing (NLP) can answer common legal questions, provide procedural guidance, and even suggest the types of documents needed for various court filings. This makes the system more user-friendly for non-experts.

- Guided Legal Aid: AI can streamline the process of applying for legal aid by automating the initial screening of applications, determining eligibility, and providing initial support, thus reducing wait times for assistance.

9. Challenges and Limitations of AI in the Judiciary

While AI brings numerous benefits, its integration into the judiciary is not without challenges:

- Algorithmic Bias: AI systems learn from historical data, and if that data includes biases, AI can perpetuate or even amplify these biases in its predictions and analyses. Ensuring unbiased data and thorough oversight of AI models is essential.

- Data Privacy Concerns: The use of AI involves processing large volumes of sensitive data, raising significant privacy and data security concerns. Proper safeguards and adherence to data protection regulations are necessary to maintain public trust.

- Reliance on Human Oversight: AI should complement, not replace, human judgment. Judges must understand the workings and limitations of AI tools to use them effectively and responsibly.

- Technological Disparities: Not all courts in India are equipped with the technological infrastructure needed for the effective implementation of AI. Bridging the digital divide between urban and rural courts is essential to ensure the benefits of AI reach all parts of the judiciary.

10. Future Prospects of AI in the Indian Judiciary

The future of AI in the Indian judiciary is promising, with potential developments including:

- Advanced Predictive Modeling: AI systems can evolve to provide even more accurate predictions by incorporating real-time legal developments and court rulings into their models.

- Integration with Blockchain: Coupling AI with blockchain technology can enhance data security and transparency in handling court records and evidence management.

- AI for Sentencing Assistance: While contentious, AI can be developed to assist judges in sentencing by analyzing case facts, comparing them with precedents, and suggesting fair sentences. This could help standardize sentencing and reduce disparities.

- AI-Enhanced Legal Databases: AI-driven enhancements to legal databases can make it easier to conduct comprehensive research that is not only fast but contextually aware, improving the quality of legal arguments and judgments.

Conclusion

AI is poised to play an increasingly influential role in the Indian judiciary. It has already begun transforming how cases are managed, researched, and resolved, making the system more efficient and accessible. However, careful consideration of ethical, regulatory, and infrastructural challenges is critical to ensuring that AI complements human judgment and upholds the principles of fairness and transparency. With the right policies and practices, AI can be a powerful ally in making the Indian judiciary a more effective and responsive institution.

KEY AI TOOLS AND PROJECTS IN INDIA

The Indian judiciary has begun to leverage a range of AI tools and projects to streamline judicial processes and improve overall efficiency. These efforts are aimed at addressing long-standing issues such as case backlogs, time-consuming legal research, and enhancing public access to justice. Below is a comprehensive overview of some of the significant AI tools and projects currently being utilized or developed within the Indian judicial system:

1. SUPACE (Supreme Court Portal for Assistance in Court Efficiency)

SUPACE, launched by the Supreme Court of India, represents a significant stride towards integrating AI into judicial operations:

- Purpose and Functionality: SUPACE is an AI-driven platform designed to assist judges by simplifying the research process. It helps in sifting through vast amounts of legal documents, case laws, and precedents to identify relevant information that supports decision-making.

- Efficiency Gains: By automating part of the research workload, SUPACE allows judges to save valuable time and focus more on the qualitative aspects of judgments, such as legal reasoning and writing.

- Impact: Though SUPACE is still in the adoption phase, it has the potential to reduce the time spent on case analysis, which can help expedite the resolution of cases.

2. AI-Powered E-Courts System

The E-Courts project is a nationwide initiative aimed at digitizing court operations. AI plays a crucial role in enhancing the capabilities of this system:

- Features: The E-Courts system uses AI to automate processes like case filing, scheduling, and document management. Additionally, machine learning algorithms help in the categorization of cases and the automation of routine tasks.

- Virtual Hearings: During the COVID-19 pandemic, the judiciary utilized AI-driven solutions to facilitate virtual hearings, enabling courts to continue functioning without significant interruptions.

- Case Status and Alerts: AI tools in the E-Courts project provide real-time updates on case statuses and automated notifications for litigants and lawyers, ensuring greater transparency and communication.

3. National Judicial Data Grid (NJDG)

The NJDG is an essential tool that helps monitor case data across courts in India:

- Data Analytics and Insights: NJDG incorporates data analytics to provide insights into case trends, backlogs, and case disposal rates. AI is used to process and analyze this large dataset, helping identify patterns that inform policy decisions and resource allocation.

- Public Access: By allowing public access to information about case pendency and court performance, NJDG enhances transparency and accountability within the judicial system.

- AI Integration: Advanced machine learning models integrated with NJDG can predict case durations and highlight courts with the highest pendency, aiding in targeted resource deployment and administrative planning.

4. Judy AI Project

The Judy AI initiative is an AI-powered legal research tool that supports lawyers and judges by providing intelligent case law search capabilities:

- AI-Driven Search: Judy AI employs natural language processing (NLP) to enable users to search legal databases using conversational language. This eliminates the need for exact keyword searches and makes research more intuitive and faster.

- Case Summaries: The tool can also summarize large volumes of legal texts, extracting relevant details and key points that assist in legal analysis.

- Ease of Use: By simplifying complex research processes, Judy AI helps lawyers and judges stay up-to-date with precedents and legal changes, ultimately contributing to more informed judgments.

5. AI Translation Tools for Multilingual Accessibility

Given the linguistic diversity of India, AI-driven translation tools play a vital role in making legal documents and proceedings more accessible:

- Automatic Translations: These AI tools can translate legal documents and judgments into multiple Indian languages. This helps litigants, especially those in rural and non-English-speaking areas, understand court proceedings and judgments.

- Speech Recognition and Transcription: AI-powered speech recognition tools are used for transcribing court proceedings in real time, which aids in maintaining accurate records and supports accessibility for individuals with hearing or visual impairments.

- Future Potential: Ongoing projects aim to enhance the accuracy and capabilities of these tools to accommodate legal jargon and contextual nuances, improving overall inclusivity in the legal system.

6. Online Dispute Resolution (ODR) Platforms

ODR platforms, supported by AI, provide a means for resolving disputes outside the traditional courtroom:

- AI-Assisted Negotiations: These platforms use AI to guide parties through the dispute resolution process, suggesting settlement terms based on case data and previous similar disputes.

- Efficiency and Cost-Effectiveness: By reducing the need for physical court appearances, ODR platforms help alleviate court workloads and offer an economical solution for litigants, especially in civil cases and smaller claims.

- Integration with Legal Ecosystem: As the popularity of ODR grows, it is being integrated into the broader legal ecosystem, with AI playing a crucial role in automating the facilitation of negotiation, arbitration, and mediation processes.

7. Virtual Legal Assistants and Chatbots

AI-powered chatbots and virtual legal assistants are becoming more common in the Indian legal system to enhance user engagement and provide basic legal assistance:

- Public Guidance: AI chatbots are used on judicial websites to help the public find information related to court procedures, case statuses, and legal documents. This reduces the burden on court clerks and improves service delivery.

- Legal Aid Support: These virtual assistants can screen legal aid applications, provide guidance on necessary documents, and direct users to appropriate legal resources, making it easier for individuals to access justice.

Challenges and Future Prospects

While the use of AI in the Indian judiciary is promising, there are certain challenges and future considerations to address:

- Ethical Concerns: Ensuring that AI systems used in the judiciary are transparent, unbiased, and accountable is a primary concern. The judiciary must develop standards and guidelines to prevent biases in AI decision-making.

- Data Privacy: Handling sensitive court data through AI platforms raises privacy and security issues that need to be managed with robust data protection frameworks.

- Training and Adaptation: Adequate training for judges, lawyers, and court staff is essential to harness the full potential of AI tools. Resistance to adopting new technology can slow down progress, so efforts must focus on building digital literacy within the judiciary.

Conclusion

The integration of AI in the Indian judiciary has already shown considerable promise in enhancing the efficiency and accessibility of legal processes. Initiatives like SUPACE, the E-Courts system, NJDG, Judy AI, and ODR platforms signify a forward-thinking approach to modernizing the legal system. However, to ensure that these technologies reach their full potential, it is critical to address associated challenges and continue fostering an environment where AI complements human judgment, ultimately making the justice system more effective and equitable.

BENEFITS OF AI IN THE INDIAN JUDICIARY

The integration of Artificial Intelligence (AI) into the Indian judiciary is driving significant transformations, enhancing the system's ability to administer justice efficiently and transparently. The benefits extend across various aspects of judicial operations, from streamlining case management to making legal information more accessible to the public. Below is an in-depth look at the key benefits AI brings to the Indian judiciary:

1. Reduction of Case Backlogs and Faster Case Resolution

One of the most pressing challenges in the Indian judicial system is the large volume of pending cases. AI offers significant advantages in addressing this issue:

- Automated Case Allocation: AI algorithms can assist in the equitable distribution of cases based on the workload and specialization of judges. This reduces delays in case handling and ensures that cases are heard by judges best equipped to deal with them.

- Case Prioritization: AI tools can help prioritize cases based on urgency and complexity. For instance, cases involving immediate human rights concerns or those with time-sensitive implications can be brought to the forefront for quicker resolution.

- Efficient Data Handling: By processing vast amounts of data in a fraction of the time that manual research would take, AI helps expedite the decision-making process. This enables courts to resolve cases more quickly and reduces the overall case backlog.

2. Enhanced Legal Research Capabilities

AI has transformed how legal professionals approach research, making it faster and more accurate:

- Quick Retrieval of Relevant Information: AI-powered legal research tools can scan through extensive legal databases, judgments, and precedents to retrieve relevant information instantaneously. This reduces the hours lawyers and judges spend on manual research.

- Contextual Search: Unlike traditional keyword-based searches, AI systems use natural language processing (NLP) to understand the context of queries, providing results that are more aligned with the user's intent.

- Error Reduction: The comprehensive and automated nature of AI research tools minimizes human error, ensuring that legal professionals do not overlook critical case laws or statutes during their research.

3. Improved Judicial Decision-Making

AI supports judges in making more informed and consistent decisions:

- Access to Precedents: AI tools can present relevant past judgments and legal principles that support or contradict certain positions, assisting judges in crafting well-reasoned judgments.

- Predictive Analysis: By analyzing historical data, AI can provide predictive insights on the likely outcomes of similar cases. This helps judges foresee potential legal implications and make decisions that are more consistent and aligned with established jurisprudence.

- Reduction in Cognitive Load: AI helps lighten the cognitive load by automating routine research and analysis tasks, allowing judges to dedicate more mental resources to interpreting the law and writing judgments.

4. Greater Accessibility and Inclusivity

AI-driven solutions are instrumental in making the judicial system more accessible to all segments of society:

- Language Translation: India's diverse linguistic landscape often creates barriers for non-English speakers. AI-powered translation tools help bridge this gap by translating judgments, legal documents, and proceedings into multiple regional languages.

- Accessibility for Differently-Abled Individuals: AI tools that convert speech to text or text to speech make legal processes more accessible for people with hearing or visual impairments. This inclusive approach ensures that more individuals can engage with the legal system effectively.

- Guided Legal Information: Chatbots and virtual assistants, powered by AI, provide legal information and procedural guidance to the general public. These tools can answer common questions and help individuals navigate complex legal procedures without needing a lawyer.

5. Efficient Management of Court Records

Managing court records can be a labor-intensive and error-prone process, but AI offers solutions that streamline and secure data handling:

- Automated Data Entry and Retrieval: AI systems can automate the entry and organization of case files, ensuring that court records are updated in real-time and are easily retrievable when needed.

- Document Summarization: AI can summarize long case documents and extract essential points, facilitating quicker understanding for judges, lawyers, and court staff.

- Enhanced Security: By using AI in conjunction with technologies like blockchain, the judiciary can ensure that court records are tamper-proof and maintain a high level of data integrity.

6. Support for Virtual Court Proceedings

The advent of virtual court proceedings, particularly during the COVID-19 pandemic, has shown the importance of AI in maintaining judicial operations remotely:

- Automated Scheduling and Notifications: AI assists in scheduling virtual hearings, sending automated reminders to involved parties, and ensuring that hearings are conducted as planned.

- Real-Time Transcription: AI-driven speech-to-text tools provide live transcriptions of court proceedings, enhancing the accuracy of records and aiding participants who require text-based documentation.

- Seamless Document Sharing: AI aids in the organization and sharing of documents during virtual hearings, reducing administrative burdens and facilitating smoother proceedings.

7. Facilitating Alternative Dispute Resolution (ADR)

AI is also making strides in promoting out-of-court settlements through online dispute resolution (ODR) platforms:

- AI-Assisted Mediation: AI can assist mediators by analyzing case facts, suggesting potential resolutions, and facilitating dialogue between disputing parties. This helps parties reach settlements faster and reduces the load on formal courts.

- Improved Efficiency: ODR platforms with integrated AI systems provide a streamlined and user-friendly way to handle smaller disputes that don't require a full court trial. This frees up court resources for more complex cases.

8. Enhanced Public Trust and Transparency

Transparency is a cornerstone of a fair judicial system, and AI contributes by:

- Improving Accountability: AI-powered data analytics provide insights into court operations, allowing stakeholders to understand trends in case handling, case durations, and verdict consistency.

- Public Access to Case Data: Platforms like the National Judicial Data Grid (NJDG), supported by AI, enable the public to access case data and track the status of cases, promoting transparency and trust in the system.

- Reducing Human Bias: While AI is not free from bias, it can help minimize human prejudices in decision-making by relying on data-driven insights and standardized procedures.

9. Cost Savings and Resource Efficiency

AI implementation can lead to significant cost savings for the judiciary and legal practitioners:

- Reduced Administrative Costs: Automation of routine tasks, such as case scheduling, document filing, and case management, lowers the need for extensive clerical work, reducing operational costs.

- Streamlined Court Operations: By reducing the time needed for research and data handling, AI allows court personnel and lawyers to focus on higher-value tasks, thereby optimizing resource allocation.

Conclusion

AI's integration into the Indian judiciary has already shown significant benefits in terms of efficiency, accessibility, transparency, and consistency. While challenges such as data privacy, bias mitigation, and infrastructure constraints remain, the potential for AI to continue transforming the legal landscape is clear. With proper regulation, continuous training for legal professionals, and ongoing technological advancements, AI can become a critical tool that aids in delivering justice more swiftly and equitably across India.

CHALLENGES AND CONCERNS

While the integration of Artificial Intelligence (AI) into the Indian judiciary system holds considerable promise, it also presents significant challenges and concerns that must be addressed to ensure its effective and ethical use. These challenges span technical, legal, ethical, and infrastructural aspects, each of which could impact the overall implementation and reliability of AI in judicial processes.

1. Data Privacy and Security

One of the foremost concerns in integrating AI into the judiciary is ensuring the privacy and security of sensitive legal data:

- Confidentiality Risks: Judicial data often contains highly confidential information. Using AI systems to process and store this data raises questions about how to safeguard it from unauthorized access, cyber-attacks, or breaches. Maintaining the integrity and privacy of legal records is critical to protecting the rights of litigants.

- Data Protection Frameworks: India's data protection laws are still evolving. The absence of a comprehensive and robust data protection framework poses a challenge in ensuring that AI tools handling judicial data comply with privacy standards and prevent misuse or unauthorized sharing.

- AI System Security: Ensuring that AI platforms used by the judiciary are secured against hacking and tampering is crucial. Any breach could compromise not just data integrity but also public trust in the judicial system.

2. Bias and Fairness

AI systems are only as objective as the data and algorithms that power them, which can pose significant fairness and equity concerns:

- Algorithmic Bias: If AI systems are trained on biased data, they may produce biased outputs that reflect or even amplify existing inequalities within the judicial system. For instance, historic data used in AI training may contain racial, gender, or socioeconomic biases, leading to unfair judgments or prioritizations.

- Lack of Transparency: Many AI algorithms, especially those involving machine learning, operate as "black boxes," meaning their decision-making processes are opaque. This lack of transparency can make it difficult to understand how AI arrived at a particular conclusion, raising ethical concerns about accountability in judicial decisions.

- Addressing Discrimination: Ensuring that AI systems are programmed and updated to recognize and eliminate biased patterns is a complex task. Without regular audits and comprehensive testing, AI applications risk perpetuating unfair practices.

3. Limited Infrastructure and Digital Divide

India's vast and diverse landscape presents challenges related to digital accessibility and the necessary infrastructure for implementing AI:

- Uneven Access to Technology: The use of AI in judiciary operations requires advanced digital infrastructure, which may not be uniformly available across all regions, especially in rural or economically disadvantaged areas. This creates a digital divide, wherein only some parts of the judiciary benefit from technological advancements.

- Training and Adaptation: The successful adoption of AI tools requires substantial training for judges, lawyers, and court staff to effectively use these technologies. Without widespread digital literacy programs, there is a risk that users may not fully understand or utilize AI's potential, leading to inefficient or incorrect usage.

- Resource Constraints: Smaller and less-resourced courts may struggle to allocate funds and resources for implementing and maintaining AI technologies. This challenge emphasizes the need for government support and strategic partnerships to enable technology access across all judicial levels.

4. Ethical and Legal Concerns

Introducing AI into a field as sensitive as the judiciary raises significant ethical questions:

- Accountability Issues: If an AI tool provides incorrect or misleading information that influences a judgment, it can be difficult to assign responsibility. The current legal framework does not clearly outline who is accountable—whether it be the developers, the judiciary, or other stakeholders—when AI errors occur.

- Ethical Use of AI: Decisions related to life, liberty, or property are inherently moral and should not be based solely on computational analyses. Relying too heavily on AI for judicial decision-making can dehumanize the process and undermine the moral responsibility of judges to interpret and apply the law based on the human condition.

- Ensuring Judicial Discretion: While AI can assist in research and case management, judicial decisions should still rest in human hands. It is important to balance the use of AI as an aid without allowing it to overstep into making rulings or decisions independently.

5. Regulatory Challenges and Policy Gaps

The current regulatory environment for AI use in the Indian judiciary is still in its infancy:

- Lack of Comprehensive Policies: There is an urgent need for clear policies and legal frameworks that outline the permissible scope and limitations of AI usage in the judiciary. Without such regulations, the potential for misuse or overreach remains high.

- Standards and Certification: There is no standardized protocol for vetting and certifying AI tools used within the judiciary. Implementing a robust certification system to evaluate the reliability, fairness, and security of AI platforms is crucial to maintain trust.

- International Benchmarks: While some countries have established guidelines for ethical AI use, India is still developing its policy stance. Adopting best practices from successful global AI implementations while considering India's unique socio-legal context is essential.

6. Dependency and Loss of Human Touch

The integration of AI poses the risk of over-dependence, potentially diminishing the human touch that is crucial in delivering justice:

- Loss of Empathy: The judiciary deals with complex human issues where empathy and understanding play key roles in delivering fair judgments. AI lacks the human capacity for empathy and moral judgment, which could limit its effectiveness in contexts requiring compassion and nuanced decision-making.

- Human Oversight: There is a concern that judges and lawyers might become too reliant on AI for routine research and decision support, leading to a gradual erosion of their analytical and interpretative skills. Maintaining a balance between technological assistance and human oversight is critical for preserving judicial wisdom and adaptability.

7. Interoperability and Integration Challenges

Integrating AI with existing judicial systems and processes poses practical challenges:

- Legacy Systems: Many courts still operate on outdated software and manual record-keeping systems. Integrating AI with these legacy systems may require extensive overhauls, incurring high costs and logistical hurdles.

- Data Standardization: For AI to function effectively, data needs to be standardized across various judicial databases and case management systems. The lack of uniformity in data formats and record-keeping practices can hinder the seamless integration of AI tools.

- Technical Expertise: Implementing and maintaining AI solutions require technical expertise that may not be readily available within the current workforce. This calls for partnerships with technology firms and increased investments in technical training for court IT staff.

Conclusion

The adoption of AI in the Indian judiciary, while promising, is not without its challenges. Addressing concerns related to data privacy, algorithmic fairness, infrastructure gaps, and ethical considerations is essential for successful integration. Policymakers and judiciary leaders must work collaboratively to create a balanced regulatory environment that maximizes the benefits of AI while mitigating its risks. Continuous training, robust data protection laws, and transparent AI practices are key to harnessing AI's full potential in supporting an equitable and effective judicial system.

FUTURE PROSPECTS

The future of AI in the Indian judiciary is both promising and multifaceted. While the integration of artificial intelligence (AI) in the judicial system is in its nascent stages, the potential for further expansion and sophistication is considerable. The ongoing efforts to incorporate AI-driven solutions are expected to shape the judiciary in profound ways, enhancing its ability to deliver justice efficiently and equitably. Below, we explore the future prospects of AI in the Indian judiciary and how it might evolve in the coming years.

1. Expansion of AI-Powered Legal Research Tools

AI has already revolutionized legal research by enabling quicker access to vast amounts of case law and statutes. The future promises even more advanced AI tools that will:

- Leverage Deep Learning: Future research tools are expected to use deep learning models capable of understanding complex legal language and context more deeply. These tools will help legal professionals find relevant cases, statutes, and legal precedents with greater precision and speed.

- Integration with Predictive Analytics: AI could evolve to not only provide information but also predict case outcomes based on previous judgments and current legal trends. This could be beneficial for lawyers in strategizing and advising clients on potential case outcomes.

- Multi-Language Capabilities: Advanced AI research tools could include expanded support for India's numerous regional languages, enabling lawyers and judges across the country to conduct research in their native languages, thus promoting inclusivity and accessibility.

2. Streamlined Case Management and Decision-Making

AI has the potential to transform case management systems, making them more efficient and intelligent:

- Automated Workflow Management: Future AI-driven case management systems could automate a wider range of judicial processes, such as generating reports, setting hearing dates, and sending notifications. This would reduce administrative burdens and improve the speed of case handling.

- Intelligent Document Summarization: Enhanced AI algorithms could summarize case files and extract essential facts and arguments more accurately, enabling judges and lawyers to review cases in less time without losing critical details.

- Advanced Analytics for Sentencing and Bail Decisions: AI models could assist judges in assessing patterns in sentencing and bail decisions, thereby promoting consistency and reducing disparities. However, such systems would need to be carefully designed to avoid reinforcing existing biases.

3. Adoption of Virtual Courtrooms and Digital Hearings

The success of virtual courtrooms during the COVID-19 pandemic has laid the groundwork for future innovations:

- AI-Enhanced Virtual Platforms: Future court proceedings may integrate AI-driven transcription services that provide real-time translations and text-to-speech functionalities, allowing seamless communication between participants who speak different languages or require accessibility tools.

- Smart Scheduling Systems: AI could enhance scheduling systems to better handle the complexities of arranging hearings, particularly for cases involving multiple parties or those needing urgent attention. These systems could also factor in the availability of legal representatives and reduce scheduling conflicts.

- Hybrid Court Models: Courts might adopt a hybrid model, where AI helps determine which cases are suited for virtual hearings and which require in-person attendance based on the nature and complexity of the case.

4. Use of AI for Predictive Policing and Law Enforcement

Beyond courtroom applications, AI could play a pivotal role in law enforcement and pre-judicial processes:

- Predictive Policing: AI could assist police departments in analyzing crime patterns and predicting potential hotspots for future crimes. This could aid in preventive measures and resource allocation, helping to address crime proactively.

- Support for Investigations: AI could assist in analyzing digital evidence, automating cross-references of case data, and connecting dots that would take human investigators significantly longer to identify.

- Ethical and Privacy Safeguards: To balance the potential benefits with ethical concerns, future AI applications in policing would need strong regulations to prevent racial, economic, or social biases in data analysis and decision-making.

5. Expansion of Online Dispute Resolution (ODR)

AI's capabilities in facilitating online dispute resolution (ODR) are expected to expand:

- AI-Facilitated Negotiation: Future ODR platforms could include AI mediators that assist in negotiations between parties by suggesting compromises and flagging contentious points. These systems could make dispute resolution more accessible and affordable, reducing the number of cases that proceed to formal court trials.

- Scalability and Customization: AI can enable scalable ODR platforms that handle a wide range of disputes, from consumer grievances to commercial conflicts, offering tailored approaches that suit different types of cases and stakeholders.

- Automated Follow-Ups: Enhanced ODR platforms could provide automated follow-ups on the enforcement of settlements, ensuring compliance and notifying relevant parties of obligations and deadlines.

6. Advanced Training and Capacity Building

The adoption of AI in the judiciary will necessitate extensive training and capacity-building efforts:

- AI Literacy Programs: For the judiciary to maximize the benefits of AI, judges, lawyers, and court staff will need training on using AI tools effectively. This training should focus not just on how to use these tools but also on understanding their limitations and ethical implications.

- Partnerships with Educational Institutions: Law schools and judicial academies may partner with technology institutes to develop specialized courses on AI's application in law. This interdisciplinary approach can ensure that upcoming legal professionals are well-versed in both legal and technological perspectives.

- Continuous Learning: Given the fast-paced nature of AI advancements, ongoing education and skill development programs for current legal practitioners will be essential to keep them updated on the latest tools and methods.

7. Strengthening Regulatory and Ethical Frameworks

The growth of AI use in the judiciary necessitates the development of clear regulatory and ethical guidelines:

- Developing Comprehensive AI Policies: Future policies will need to establish guidelines for ethical AI use, addressing concerns such as data privacy, accountability, and the prevention of algorithmic bias.

- International Collaboration: India can benefit from collaborating with international organizations and other countries to adopt best practices for ethical AI use in the judiciary.

- Robust Oversight Mechanisms: Independent oversight bodies could be established to monitor AI applications, ensure adherence to ethical standards, and maintain public confidence in AI-enhanced judicial systems.

8. Integration with Blockchain for Enhanced Security

AI's future integration with blockchain technology could strengthen the judiciary's data management practices:

- Tamper-Proof Records: Combining AI with blockchain could ensure that court records and documents remain tamper-proof, enhancing the reliability of evidence and reducing the risk of data manipulation.

- Smart Contracts for Legal Agreements: AI-powered platforms using blockchain could automate the execution of legal contracts, where terms are met and enforced automatically, reducing the need for judicial intervention in contract disputes.

- Secure Data Sharing: AI systems, in conjunction with blockchain, could facilitate secure data sharing among different judicial departments and law enforcement agencies, ensuring data integrity and confidentiality.

Conclusion

NARAYANASWAMY.R

The future prospects of AI in the Indian judiciary are vast and transformative. As AI technology advances, it is likely to play a critical role in enhancing the efficiency, accessibility, and fairness of judicial processes. However, achieving these benefits will require comprehensive strategies that address ethical considerations, infrastructure challenges, and regulatory frameworks. A collaborative approach involving the judiciary, technology developers, policymakers, and legal professionals will be essential to harness AI's full potential while safeguarding the principles of justice and equity.

REGULATORY AND ETHICAL CONSIDERATIONS

The integration of Artificial Intelligence (AI) into the Indian judiciary system, while promising for improving efficiency and access to justice, comes with a significant set of regulatory and ethical challenges. Addressing these considerations is vital to ensure that AI is used responsibly, transparently, and fairly in judicial contexts. These aspects cover the creation of new laws, updating existing frameworks, establishing oversight mechanisms, and embedding ethical principles within AI development and deployment.

1. Developing Comprehensive Legal and Regulatory Frameworks

The introduction of AI in the judiciary requires a strong regulatory structure to guide its usage and address potential pitfalls:

- Establishment of AI-Specific Laws: India needs comprehensive legislation that defines the scope and permissible use of AI in the judiciary. This includes laws that outline the parameters for data use, algorithmic transparency, accountability for AI-generated decisions, and data ownership rights.

- Adapting Existing Laws: Current data protection and privacy laws, such as the Digital Personal Data Protection Act, must be updated to incorporate clauses that address AI applications specifically, focusing on the unique challenges posed by algorithmic data processing and automated decision-making.

- Policy Directives: Strategic policy directives from bodies like the Ministry of Law and Justice and the Supreme Court should ensure consistency in the deployment of AI tools across various judicial levels and states.

2. Data Privacy and Confidentiality

AI systems used in the judiciary must handle vast amounts of sensitive and personal data:

- Data Collection and Storage Standards: Regulations should mandate strict protocols for how judicial data is collected, processed, stored, and shared by AI systems to protect litigants' privacy.

- Anonymization Practices: Robust data anonymization techniques should be required to prevent the misuse of data, ensuring that individuals' identities remain protected even when data is analyzed for case trends or AI model training.

- Adherence to International Standards: Aligning with global best practices, such as those outlined by the European Union's General Data Protection Regulation (GDPR), could strengthen data privacy standards in India and protect against data breaches and misuse.

3. Accountability and Liability

Determining accountability when using AI for judicial purposes poses unique challenges:

- Clarifying Responsibility: Establishing clear guidelines for who is responsible if an AI system generates erroneous advice or influences a judicial outcome incorrectly is crucial. The division of liability between developers, operators, and the judiciary must be precisely defined.

- Human Oversight: AI tools should complement human judgment rather than replace it. Regulations should ensure that final judicial decisions remain with human judges who understand the context and nuances that AI may not fully capture. This principle, often referred to as human-in-the-loop, helps mitigate risks associated with over-reliance on AI.

- Error Rectification Mechanisms: There should be a legal provision for appealing or correcting AI-driven decisions or recommendations, emphasizing that AI's role is supportive, not definitive.

4. Algorithmic Transparency and Bias Prevention

Ensuring that AI systems are transparent and unbiased is fundamental to maintaining fairness in the judiciary:

- Transparency Protocols: Developers should be required to disclose how AI algorithms are trained and the types of data used. Such transparency helps stakeholders understand how decisions are reached and whether there are any inherent biases or limitations.

- Auditable Systems: Judicial AI applications should be subject to regular audits by independent bodies to assess whether they produce fair, unbiased, and consistent results. This oversight ensures that AI systems do not perpetuate or exacerbate existing social, gender, or racial biases.

- Inclusive Datasets: AI models should be trained on diverse datasets representing all segments of society to prevent biased outcomes. This can be especially challenging in a country as diverse as India, where socioeconomic, linguistic, and regional disparities exist.

5. Ethical Considerations in Decision-Making

AI's use in the judiciary must adhere to ethical guidelines that prioritize human rights, fairness, and impartiality:

- Moral Responsibility: AI applications in the judiciary must align with the core principles of justice, emphasizing the moral responsibility that comes with making decisions impacting individuals' lives. While AI can assist in processing data or predicting outcomes, decisions involving fundamental rights and freedoms should remain human-centric.

- Transparency in AI Decisions: Ethical guidelines should mandate that AI systems provide reasons for their recommendations. This requirement ensures that AI doesn't become a "black box" where decisions are made without clear explanations, thus upholding the right to a fair trial.

- Respect for Dignity and Autonomy: Ethical AI use in the judiciary should respect the dignity and autonomy of all individuals involved. For example, AI should not overstep its bounds by making conclusive judgments about guilt or innocence, leaving these determinations to human judges who can appreciate the complexities and subtleties of cases.

6. Public Trust and Acceptance

For AI to be effectively integrated into the judiciary, public confidence is essential:

- Transparency to Build Trust: Ensuring that AI systems are transparent about their capabilities and limitations is key to building trust. This involves public communication about how AI is used and the safeguards in place to protect against errors or biases.

- Public Participation in Policymaking: Engaging with stakeholders, including legal professionals, human rights organizations, and civil society, in the policymaking process can foster broader acceptance and understanding of AI's role in the judiciary.

- Educational Initiatives: To promote public trust, educational campaigns should be conducted to inform people about the benefits and limitations of AI in the legal system. These initiatives can dispel myths and address concerns about AI replacing human judges or diminishing the quality of justice.

7. Technological and Implementation Challenges

Implementing AI in a manner that adheres to regulatory and ethical guidelines comes with its own set of challenges:

- Standardization Issues: Courts across India have varying levels of digital maturity. Establishing uniform AI standards for implementation can be challenging, requiring tailored solutions that account for regional differences.

- Resource Allocation: Adequate funding and resources must be directed towards not only deploying AI systems but also maintaining them, training staff, and updating algorithms to reflect new regulations and ethical guidelines.

- Monitoring and Continuous Improvement: Ongoing monitoring of AI systems is essential to ensure compliance with regulatory and ethical standards. This may involve periodic updates and the refinement of algorithms based on feedback and evolving legal needs.

Conclusion

The integration of AI in the Indian judiciary comes with a spectrum of regulatory and ethical challenges that require meticulous attention. To ensure AI's effective, fair, and responsible use, it is essential to develop comprehensive legal frameworks, enhance data privacy standards, maintain human oversight, and uphold transparency and fairness in AI models. Additionally, public trust, robust accountability measures, and ethical adherence are crucial for AI's long-term role in supporting the judiciary. With careful planning and a commitment to ethical governance, India's judiciary can harness AI's capabilities to enhance justice delivery while preserving fundamental human rights and trust in the legal system.

CONCLUSION

The integration of artificial intelligence into the Indian judiciary holds transformative potential, promising significant advancements in the efficiency, accessibility, and fairness of the justice system. However, realizing these benefits requires a carefully balanced approach that addresses regulatory, ethical, and technological challenges. The future of AI in the judiciary must be navigated with a deep understanding of its capabilities and limitations, ensuring that its application aligns with the principles of justice, transparency, and accountability.

A robust regulatory framework is essential to guide the deployment of AI tools in judicial processes. This includes clear laws that establish how AI can be used, what data handling practices should be employed, and who holds liability for errors or biases. Regulations must also focus on safeguarding data privacy and ensuring that AI systems are subject to oversight, auditing, and continuous monitoring to prevent the entrenchment of biases or the misuse of sensitive information.

Ethical considerations play a crucial role in shaping AI's place within the judiciary. The introduction of AI must uphold human dignity, ensure impartial decision-making, and avoid undue influence over judgments that require human intuition and empathy. Human oversight is vital to maintaining the integrity of judicial processes; AI should serve as a supportive tool rather than an autonomous decision-maker, allowing judges to leverage its benefits without undermining their discretion.

Public trust and confidence are pivotal for the successful integration of AI into the judiciary. Transparent communication about how AI systems function, what safeguards are in place, and how decisions are made can help mitigate public apprehension. Educational initiatives and public engagement in policymaking can further promote acceptance and understanding, dispelling fears of technology replacing human judgment and emphasizing its role as a facilitator of justice.

The future of AI in the Indian judiciary is promising, with potential expansions into more sophisticated legal research tools, enhanced case management systems, and improved online dispute resolution mechanisms. However, achieving this vision requires consistent investment in resources, training for legal professionals, and collaboration between the judiciary, policymakers, and technologists. Cross-sector partnerships and adherence to international standards can contribute to creating a robust and responsible framework for AI use in law.

In conclusion, while AI can significantly enhance the capabilities of the Indian judiciary, it must be integrated thoughtfully, upholding the principles of fairness, transparency, and human oversight. This ensures that the judiciary remains not only more efficient but also more just and equitable. By addressing regulatory, ethical, and public trust challenges, India can harness

AI's transformative potential to create a judiciary that better serves the needs of its diverse population and meets the demands of a modern, technology-driven world.

www.ingramcontent.com/pod-product-compliance
Lightning Source LLC
Chambersburg PA
CBHW070132230526
45472CB00004B/1517